YOUR AMAZING MUSCLES

DWAYNE HICKS

New York

Published in 2023 by The Rosen Publishing Group, Inc.
29 East 21st Street, New York, NY 10010

First Edition

Editor: Greg Roza
Designer: Michael Flynn

Photo Credits: Cover, p.1 Sharomka/Shutterstock.com; p. 5 Anatoliy Karlyuk/Shutterstock.com; p. 7 S K Chavan/Shutterstock.com; p. 9 Dragon Images/Shutterstock.com; p. 11 La Gorda/Shutterstock.com; p. 13 SciePro/Shutterstock.com; p. 15 Explode/Shutterstock.com; p. 17 Roman Samborskyi/Shutterstock.com; p. 19 BearFotos/Shutterstock.com; p. 21 LightField Studios/Shutterstock.com.

Cataloging-in-Publication Data

Names: Hicks, Dwayne.
Title: Your amazing muscles / Dwayne Hicks.
Description: New York : Powerkids Press, 2023. | Series: Your amazing body | Includes glossary and index.
Identifiers: ISBN 9781725339699 (pbk.) | ISBN 9781725339712 (library bound) | ISBN 9781725339705 (6pack) | ISBN 9781725339729 (ebook)
Subjects: LCSH: Muscles–Juvenile literature.
Classification: LCC QP301.H54 2023 | DDC 612.7'4-dc23

Manufactured in the United States of America

CONTENTS

Lots of Muscle! 4
Skeletal Muscles 6
Rubber Bands 8
Joints and Tendons. 10
Have a Heart! 12
Smooth Muscles 14
Making Faces 16
Muscle Memory? 18
Lifting Weights 20
Glossary 22
For Further Information 23
Index 24

Lots of Muscle!

Scientists say that we have more than 650 muscles! These muscles can be found throughout the human body. They have a very important job. Without muscles, you wouldn't be able to move! Muscles also have other jobs in the body.

Skeletal Muscles

The muscles that allow you to move are called skeletal muscles. That's because they are **connected** to the bones of your **skeletal system**! These are the muscles most people think about when they hear the word "muscles."

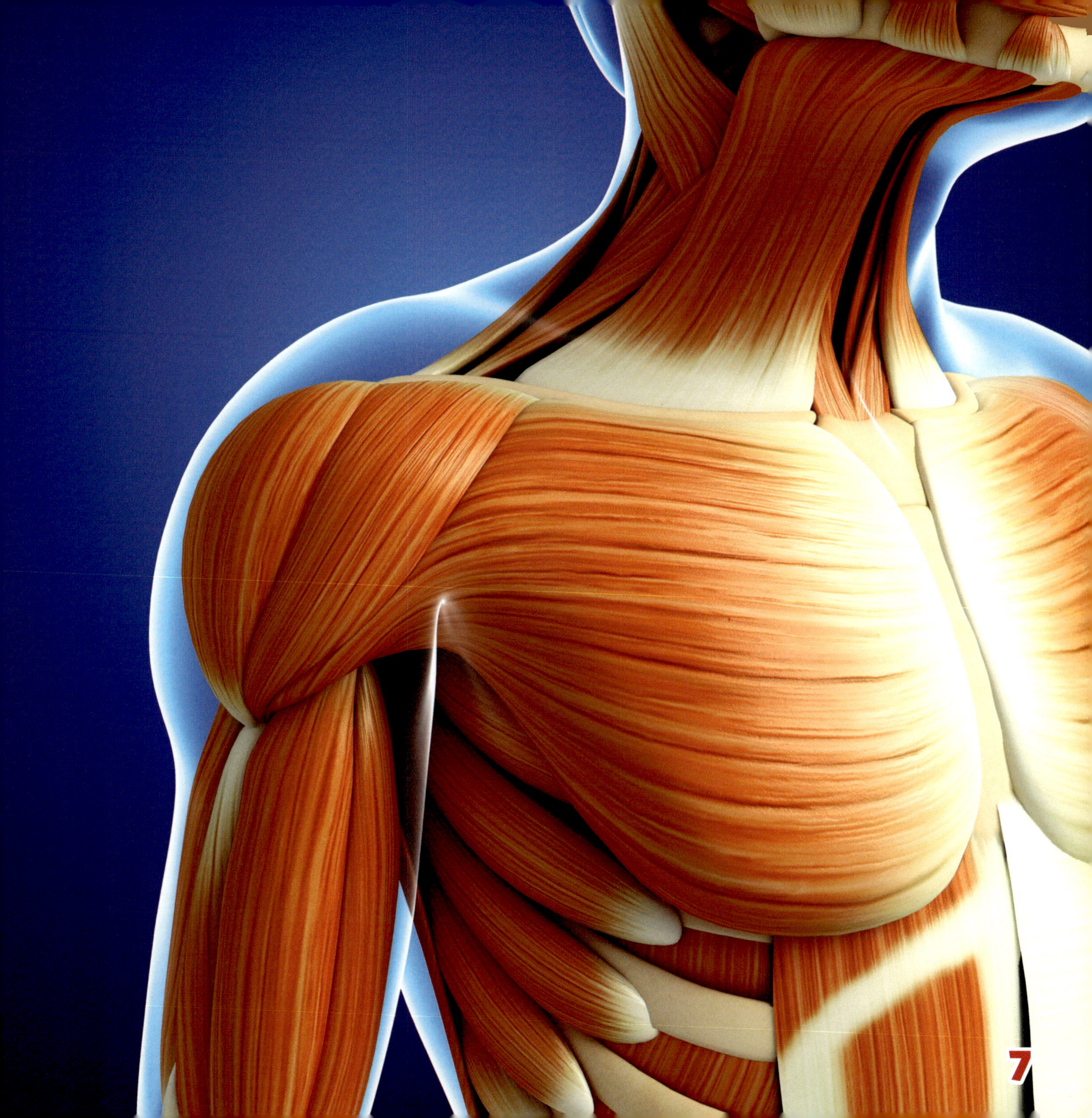

Rubber Bands

Skeletal muscles are made up of long, thin **tissues**. These tissues are like rubber bands! They expand (get bigger) and contract (get smaller). This allows us to move. The more you use your muscles, the bigger and stronger they get!

Joints and Tendons

Skeletal muscles are connected to your bones by tough tissues called tendons. The point where two bones meet is called a joint. Tendons connect a muscle to the two bones on either side of a joint. Joints and tendons help you move too!

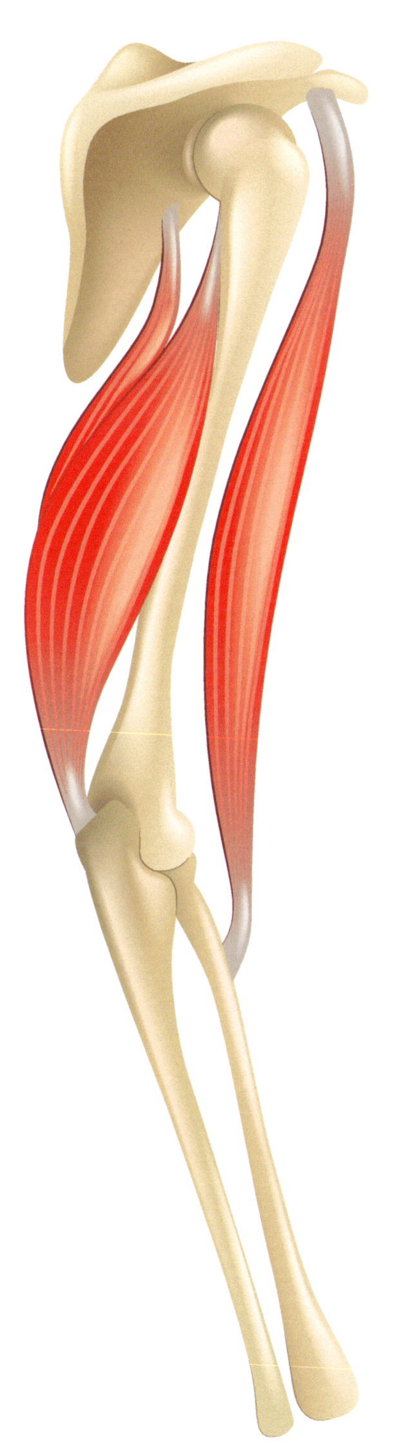

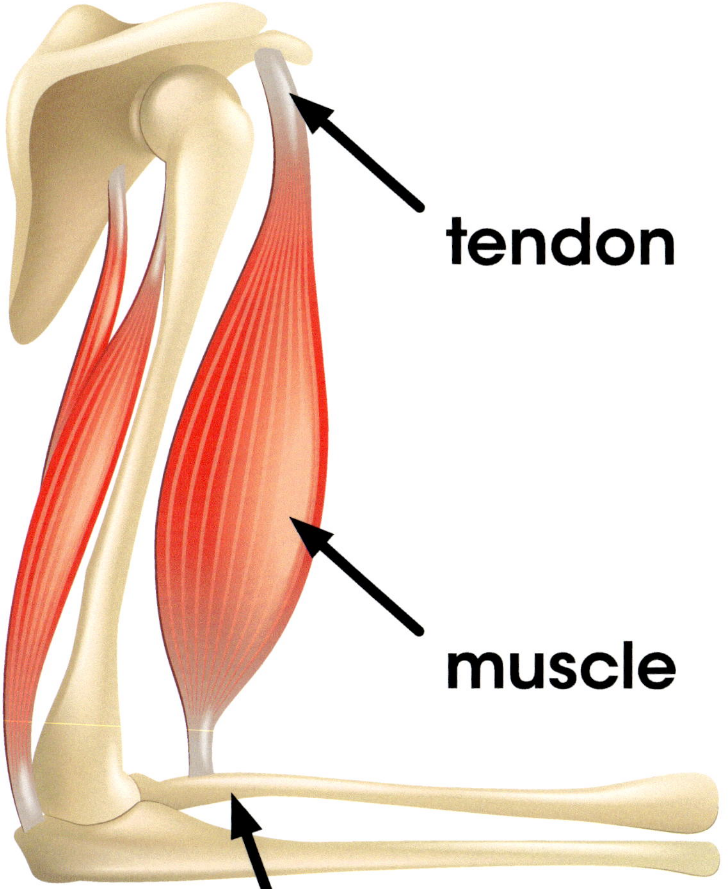

tendon

muscle

bone

Have a Heart!

You might be surprised to hear that your heart is the hardest-working muscle in your body. Your heart muscles keep your heart beating for your whole life without you even thinking about it! Heart muscle tissue is thicker than skeletal muscle tissue.

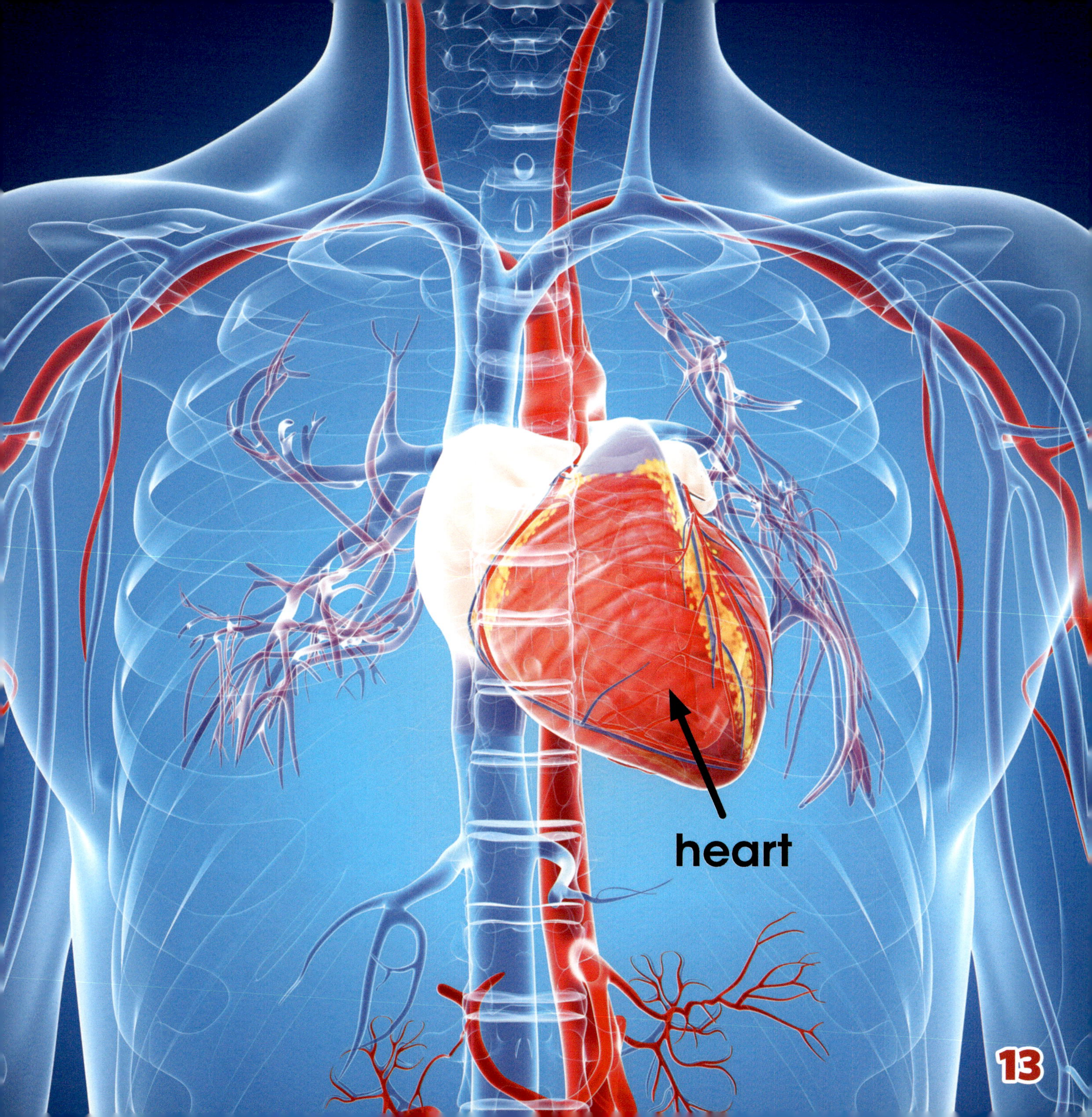
heart

Smooth Muscles

Smooth muscles line the walls of some of your body parts. Your **digestive system** uses smooth muscles. The muscles expand and contract to move food through the body. Your brain controls smooth muscles without you even knowing it.

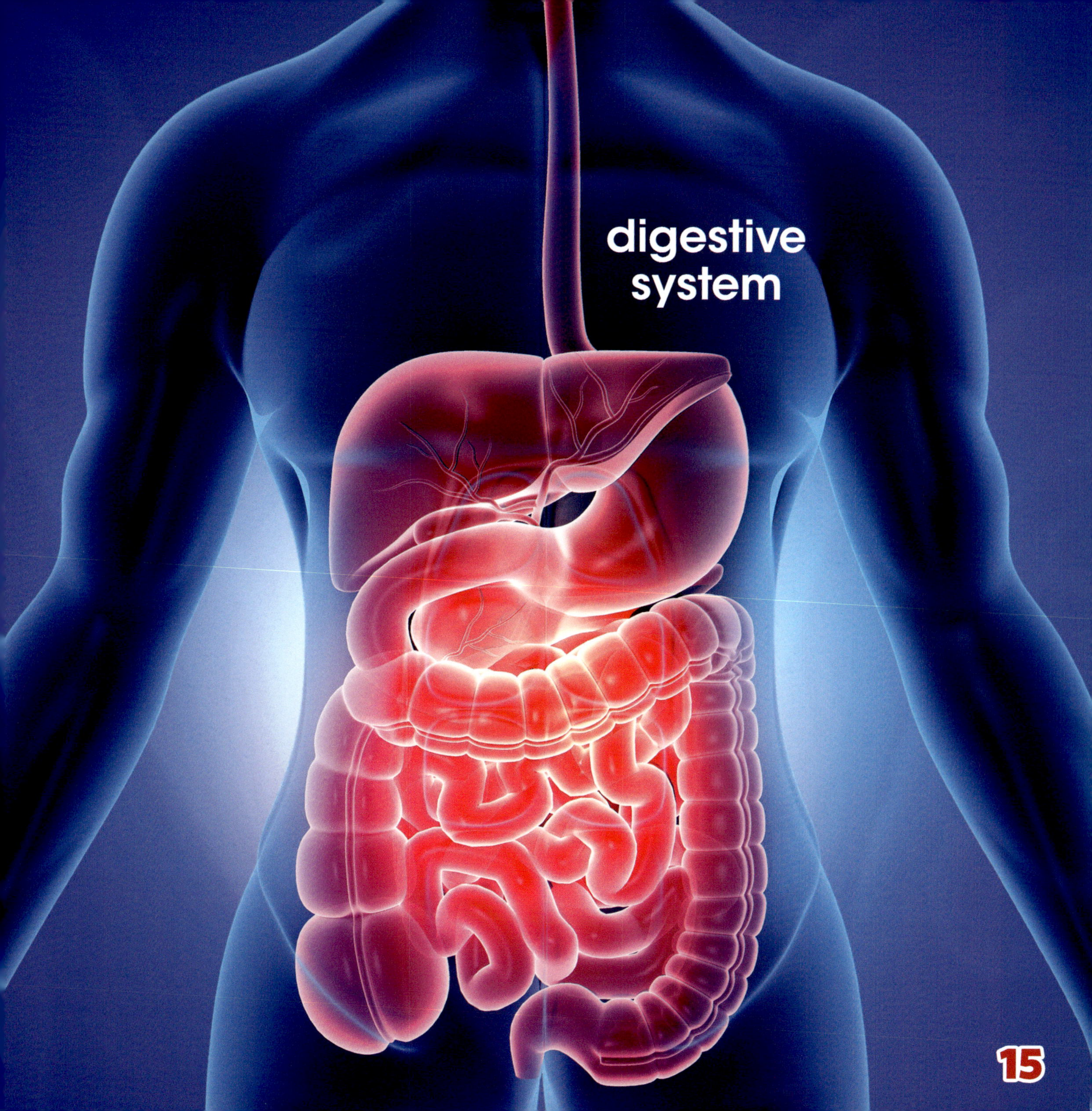
digestive
system

Making Faces

Have you ever thought about all the different faces you can make? You can look sad, happy, and even goofy! You can make all those faces because of the many skeletal muscles in your face. Your tongue is also a muscle!

Muscle Memory?

Learning a new sport or activity is hard to do at first. After you do something for a long time, it gets much easier. **Athletes** sometimes call this "muscle **memory**." However, your muscles can't remember things! It's actually your brain remembering the movements.

Lifting Weights

Using your muscles is the best way to keep them healthy! Lifting weights several times a week will help your muscles grow larger and stronger. An adult can show you how to lift weights safely. Always follow the rules to avoid muscle **injuries**.

GLOSSARY

athlete: A person who is trained in or good at games and exercises that require physical skill and strength.

connected: Joined together.

digestive system: The body parts, including the mouth and stomach, that break down food so the body can use it.

injury: Harm, damage, or loss.

memory: The store of things learned and kept in the mind.

skeletal system: The bones of the body, or the "skeleton."

tissue: A group of cells of the same kind that come together to form the basic parts that make up a plant or animal.

FOR FURTHER INFORMATION

WEBSITES

Why Exercise Is Wise
kidshealth.org/en/kids/work-it-out.html
Want to learn more about exercise and building stronger muscles? This website can get you started.

Your Muscles
kidshealth.org/en/kids/muscles.html
Read more about your muscles and view a slideshow about muscles and joints at this informative website.

BOOKS

Howel, Izzy. *Muscles and Movement.* London, UK: Wayland, 2021.

Markovics, Joyce. *Muscles.* Ann Arbor, MI: Cherry Lake Publishing, 2022.

Publisher's note to parents and teachers: Our editors have reviewed the websites listed here to make sure they're suitable for students. However, websites may change frequently. Please note that students should always be supervised when they access the internet.

INDEX

B
bones, 6, 10
brain, 14, 18

D
digestive system, 14

F
face, 16

H
heart muscle, 12

J
joints, 10

L
lifting weights, 20

M
muscle injuries, 20
muscle memory, 18

S
skeletal muscles, 6, 8, 10, 12, 16
skeletal system, 6
smooth muscle, 14

T
tendons, 10
tissues, 8, 10, 12
tongue, 16